# MINUTE MATH®

*Bothell, WA • Chicago, IL • Columbus, OH • New York, NY*

www.everydaymath.com

Send all inquiries to:
McGraw-Hill Education
8787 Orion Place
Columbus, OH 43240

ISBN: 978-0-02-141412-3
MHID: 0-02-141412-2

Printed in Mexico.

3 4 5 6 7 8 9 DRY 19 18 17 16 15 14

**Authors**
Jean Bell
Max Bell
Dorothy Freedman
Nancy Goodsell (First Edition)
Nancy Hanvey
Kate Morrison

**Contributors**
Kathryn Flores
Mary Fullmer
Rosalie Fruchter
Allison M. Greer
Deborah Arron Leslie
Curtis Lieneck
Amanda Louise Ruch
Elizabet Spaepen

# Contents

## Introduction

*Kindergarten Minute Math*® is an important part of the *Kindergarten Everyday Mathematics*® program. It reinforces the activities in the *Teacher's Lesson Guide*™, with special emphasis on developing problem-solving and mental mathematics skills. Plan to use this book regularly. Because *Minute Math* activities are short and require no (or very few) materials, they are easy to incorporate throughout the day (meetings, lining up, gathering for group time, waiting times, and so on). You might hang your *Minute Math* book on a hook in your classroom meeting area or near the doorway so you can find it easily when the opportunity for a quick activity presents itself. Some teachers even carry the book in their pocket so it is always available!

Although the book is divided into the following four parts, activities can be used with appropriate modifications at any time and recycled often throughout the year. We also encourage you and your class to invent your own *Minute Math* activities.

Routines: Pages 1–12 provide ideas for reinforcing time concepts as introduced in the Daily Routines.

Part 1: Pages 13–78 support the Daily Routines and Sections 1–3 in the *Teacher's Lesson Guide.*

Part 2: Pages 79–184 support the Daily Routines and Sections 4–6 in the *Teacher's Lesson Guide.*

Part 3: Pages 185–258 support the Daily Routines and Sections 7–9 in the *Teacher's Lesson Guide.*

## Routines

The activities in the Routines section support and reinforce the Daily Routines in the *Teacher's Lesson Guide* and other informal foundational work with the measurement of time, including seasons, months of the year, days of the week, times of day, and the use of clocks, calendars, and other tools. Remember that these activities can be varied and repeated as often as you like throughout the year.

# Notes

## Expressions of Time

Expressions such as *before and after, early and late, yesterday and tomorrow,* and *next year and last year* often confuse young children. As you pose questions such as those below and talk with children about time concepts, be alert to children having difficulty with these and other terms that describe time periods.

Ask children:

- *When do you come to school?* (In the morning)
- *When do you eat lunch?* (At noon, in the middle of the day)
- *When do most people sleep?* (At night)
- *When will you come back to school?* (Tomorrow, Monday)
- *When did we take our walk?* (Yesterday)
- *What is the coldest time of year?* (Winter)
- *When do plants outside begin to turn green?* (Spring)

CCSS **SMP7**

## Days of the Week

Ask children:

*What day of the week is it today?*

*What day was it yesterday?*

*What day will it be tomorrow?*

*What day will it be after that?*

## How Many Days?

Ask: *How many days are in a week?* (7 days)

*How many days are in a weekend?* (2 days) *Which days are they?* (Saturday and Sunday)

*How many days per week do we usually come to school?* (5 days) *Which days are they?* (Monday, Tuesday, Wednesday, Thursday, and Friday)

## Keeping Track of Time

Ask: *What are some things in the room that help us keep track of time?* (Clock, kitchen timer, teacher's watch)

*What other items can you think of that help us keep track of time?* (Calendar for days, weeks, months, year; Daily Schedule for days; and so on)

 CCSS SMP5

## Months and Seasons

Ask children:

*What season is this?*

*What month is it now?*

*What was last month?*

*What will next month be?*

*What month will it be after that?*

*What will the next season be?*

CCSS SMP7

## Birthday Card

### Number Stories

Anna's grandmother had a birthday on November 13.
(Use any recent date.)

Anna mailed her a card 3 days before her birthday. On what date did she mail it? (November 10)

It took her card 3 days to be delivered. On what date was it delivered? (November 13) (Use a calendar to help children find the correct date.)



CCSS K.CC.1, K.CC.2, K.OA.2, SMP1, SMP4, SMP5

## Using the Growing Number Line

Have children look at the Growing Number Line (Routine 1). Ask questions such as:

*What number day is it today? How many days have we been in school?*

*What number day was it yesterday?*

*What number day will it be tomorrow? Ten days from now?*

CCSS **K.CC.1, K.CC.2, K.CC.3, SMP5**

Routines

## Attendance

Ask the following questions after children complete the daily Attendance Routine (Routine 2).

*How many children are present today?*

*How many children are absent today?*

*How many children were present yesterday?*

*Are there more or fewer children present today than yesterday?*



CCSS **K.CC.5, K.CC.6, K.CC.7, SMP4**

## Counting Days in School

Invite children to look at the Concrete Number Count Collection (Routine 1). Ask:

*How many days have we been in school?*

*How many sticks (or straws) do we have?*

*How many groups of ten do we have? How many is that all together?*

*How many ones (single or leftover sticks) do we have?*

CCSS **K.CC.1, K.CC.3, K.CC.5, K.NBT.1**

## Temperature

Direct children to look at your temperature display (Routine 4). Ask:

*What is the color zone of the temperature today? What was the color zone of the temperature yesterday? Can we tell if it is warmer or cooler today than yesterday?*

*What do you predict the color zone of the temperature will be tomorrow? What about next month? What about next season?*



CCSS **SMP5, SMP7**

# Part 1

The activities in Part 1 reinforce the Daily Routines and the lessons from Sections 1–3 of the *Teacher's Lesson Guide*. Remember that any *Minute Math* activity can be varied and used as often as you like throughout the year.

# Notes

# Introduction to Counting

## Counting and Cardinality

Always be on the lookout for opportunities to ask questions that require counting, such as:

*How many children are present today?*

*How many children are buying lunch today?*

*How many children brought their own lunches today?*

*How many more children brought their lunch than are buying lunch today?*

CCSS K.CC.1, K.CC.4, K.CC.5, K.CC.7

## Finger Talk

### Counting and Cardinality

Ask: *Can fingers talk?*

*Can you use 1 finger to say "Shhhh"?* (Hold 1 finger to lips; have children copy.)

*Can you use 2 fingers to say "Think hard"?* (Tap forehead with 2 fingers.)

*Can you use 3 fingers to say "Oops! I said that wrong"?* (Cover lips with 3 fingers.)

*Can you use 4 fingers to say "Goodbye"?* (Waggle 4 fingers.)

*What can you use 5 fingers for?* (Waving hello, giving a high-five)



CCSS K.CC.4, K.CC.5

## Guessing Shapes Drawn on Backs

### Geometry

Have partners use fingers to take turns drawing shapes on each other's backs. Have children guess the shape and find an example of that shape in the classroom.

If some children need added practice in "feeling" shapes, have their partners identify the shapes as they draw them.

CCSS K.G.1, K.G.2, K.G.5, SMP4

# Counting to 10

## Counting and Cardinality

Say: *Show 10 fingers if you can count to 10.*

Call on individual volunteers to count out loud to 10. Then have the class count to 10 in unison. You may wish to have children wiggle a finger for each number they say as they say it.



CCSS **K.CC.1, K.CC.4, K.CC.5**

# Introduction to Number Stories

## Operations and Algebraic Thinking

Number Stories

Children will find number stories more interesting if you use their names and situations with which they are familiar. For example: *James, Rita, and Marcus were the only kids having a snack. How many kids were at the snack table?*

Encourage children to get in the habit of using numbers and naming the objects being counted—the units (3 cats, 5 crackers, and so on).

Whenever possible, have children act out number stories, especially ones they have difficulty understanding. Later in the year, display diagrams or number sentences to keep track of the action.

CCSS K.OA.1, K.OA.2, SMP1, SMP2

## Outdoor Fun

### Operations and Algebraic Thinking Number Stories

A crow, a robin, a blue jay, and a cardinal were sitting on a fence. How many birds were sitting on the fence? (4 birds)

Bridget was hunting for treasures on the beach. She found a shell, a feather, and a stone. How many things did Bridget find? (3 things)

Kathleen, the twins, and Leo were having a picnic. How many children were having a picnic? (4 children)



CCSS K.CC.4, K.CC.5, K.OA.1 K.OA.2, SMP1, SMP2

# What Numbers Have You Seen?

## Counting and Cardinality

Ask: *From the time you woke up this morning until now, what numbers have you seen?*

*When and how have you used numbers today?* (Counting snacks, looking at the clock or calendar, seeing the room number, and so on)

CCSS **K.CC.3, SMP2**

## Did I Do It Right?

### Counting and Cardinality

Say: *I'm going to count to 5: 1, 2, 4, 3, 5. Did I do it right?* (No.) *What did I do wrong?* (Switched 3 and 4)

*How would you count to 5?* (Ask individual children.) *Let's do it together: 1, 2, 3, 4, 5.*



CCSS **K.CC.1, K.CC.4a, SMP6**

# Listen Carefully

## Operations and Algebraic Thinking Number Stories

Shane was going to visit his grandmother. He packed his teddy bear, a box of cereal, his baseball cap, his baseball cards, and a box of crayons. How many things to wear did Shane pack? (1 thing to wear)

Shane's dog is 5 years old. His cat is 6. Which pet is younger? (Shane's dog)

Shane has had his dog for 5 years. The cat has lived with them for 4 years. Which animal has lived with them longer? (His dog)

(Display the numbers children are comparing if possible.)

CCSS K.CC.5, K.CC.7, K.OA.1, K.OA.2, SMP1, SMP2

# Clap and Count

## Counting and Cardinality

Have children listen and count as you slowly clap or stomp to a particular number. Call on volunteers to say the total number of claps or stomps.

Alternatively, a child may act as the leader for this activity.



CCSS K.CC.4, K.CC.5

## Concept of Zero

### Counting and Cardinality

Ask: *How many live giraffes are there in this room?* Reiterate that zero is the number for none, and have children write the number 0 in the air using big arm motions.

Have children "sky-write" their answers to questions such as the following, interspersing some whose answers are zero:

*How many windows are there in this room? How many doors? How many fireplaces?*

*How many clocks are in the room? How many whiteboards? How many bookshelves? How many tables? How many real dinosaurs?*

CCSS **K.CC.3, K.CC.4, K.CC.5, SMP2**

# Describe Number Shapes

## Counting and Cardinality

Describe a number's shape and have children guess which number it is. For example: *I am thinking of a number. It looks like two circles hooked together, one on top of the other. What number is it?* (The number 8)

*I'm thinking of a number. It has one straight line at the top and another straight line going down. What number is it?* (The number 7)

Children can "finger-write" each number in the palms of their hands as it is described.

After a few examples, invite a volunteer to describe a number for the class.



CCSS **K.CC.3, SMP2**

## Counting and Moving

### Counting and Cardinality

Keep counting fun by counting in conjunction with body movements:

*1, 2,* clap

*3, 4,* leg slap

*5, 6,* foot stamp

*7, 8,* head knock

*9, 10,* chest pound

CCSS K.CC.1

# Different Types

## Operations and Algebraic Thinking Number Stories

Noah asked the children at his table to name their favorite color. The answers were red, blue, pink, green, and blue. How many *different* colors were named? (4 different colors)

Michael's mom likes to swim laps. She swims freestyle, backstroke, and butterfly. How many different ways does she swim? (3 ways)



CCSS K.CC.4, K.CC.5, K.OA.1, K.OA.2, SMP1, SMP2

## Rhythmic Clapping

### Counting and Cardinality

Clap out a rhythmic pattern. Have children listen and clap back your pattern, individually or in unison.

Count the number of claps together.

CCSS K.CC.4, K.CC.5, SMP7

## Animal Sounds

### Counting and Cardinality

Whisper a number to one child. Have the child make an animal sound (moo, chirp, woof, quack, and so on) the given number of times while the other children count silently and keep track. Everyone can say the number at the end.



CCSS K.CC.4, K.CC.5

# Horses

## Operations and Algebraic Thinking **Number Stories**

Shannon collects toy horses. She had 4 horses. Her brother gave her 2 more. How many horses does she have now? (6 horses) (Children can use their fingers to show the number of horses.)

She also owns some books about horses. She got 3 more books for her birthday. Now how many horse books does Shannon have? (We can't tell because we don't know how many she started with.)

CCSS **K.OA.1, K.OA.2, SMP1, SMP2, SMP4, SMP6**

## "Noodle Knocks"

### Counting and Cardinality

Pose various kinds of "knocking" counts:

*Count and do 9 "Noodle Knocks" with me.* (Knock on your head with your knuckles.)

*Count and do 11 "Knee Knocks" with me.*

Repeat with different numbers and types of "knocks."



CCSS K.CC.4, K.CC.5

## "Simon Says"

### Counting and Cardinality

Play *Simon Says* with children, using different numbers of actions:

*Simon says, "Tap your nose 10 times."*

*Simon says, "Nod your head 7 times."*

*Now shrug your shoulders 6 times. (Don't shrug—Simon didn't say!)*

Repeat with different numbers and movements.

CCSS **K.CC.4, K.CC.5**

# More Clap and Count

## Counting and Cardinality

Have children clap or stomp to a number you show them—one clap or stomp for each count. Give the number by holding up a number card or a number of fingers, pointing to a number on the Growing Number Line, or writing a number on a slate.



CCSS K.CC.3, K.CC.4, K.CC.5, SMP2

## Tap Your Tummy

### Counting and Cardinality

Say: *Tap your tummy and count with me to 6* (or *10,* or any number)—*one tap for each count.*

CCSS **K.CC.1, K.CC.4, K.CC.5, SMP2**

# Counting Backward

## Counting and Cardinality

Say: *Let's count backward. We'll start with 10 and go back to 0. Do it with me. 10, 9, 8, 7, . . .*

You can point to the Growing Number Line as you count.



CCSS K.CC.1, K.CC.3

## Chipmunks and Cookies

### Operations and Algebraic Thinking Number Stories

Children can use their fingers to show the numbers of objects in each story. When children respond, remind them to say what the numbers describe—the units (chipmunks, cookies, and so on).

*There were 7 baby chipmunks playing beside a tree. Then 2 of the chipmunks went inside a hole in the tree. How many chipmunks are playing beside the tree now?* (5 chipmunks)

*Karen has 2 cookies. Her brother gave her 2 more cookies. How many cookies does Karen have now?* (4 cookies)

CCSS K.OA.1, K.OA.2, SMP2, SMP4

## Countdown

### Counting and Cardinality

Use the "Rocket Blast-Off Countdown" chant (*10, 9, 8, 7, 6, 5, 4, 3, 2, 1, 0, BLAST OFF!*) when transitioning between activities in the classroom or on the playground.



CCSS K.CC.1

## Missing Number

### Counting and Cardinality

Count aloud very slowly and leave out one number. Ask: *Which number have I left out?*

*1, 2, 4, 5, 6* (3)

*7, 9, 10, 11, 12* (8)

*4, 5, 6, 8, 9* (7)

*10, 8, 7, 6* (9)

If children have difficulty, help them by saying:

*1, 2, blank, 4, 5, 6*

*7, blank, 9, 10, 11, 12*

CCSS **K.CC.1, K.CC.2, SMP6**

# Show and Tell

## Operations and Algebraic Thinking

**Number Stories**

Some children brought their favorite stuffed animals to class for Show and Tell. Lily brought her large, brown bear. Devante brought his black and white spotted dog. Tiffany brought her pink cat. Matt brought his small, light brown bear.

How many stuffed animals came to Show and Tell? (4 stuffed animals)

How many bears came to Show and Tell? (2 bears)

How many boys brought bears? (1 boy)

How many children brought animals? (4 children)



CCSS **K.CC.4, K.CC.5, K.OA.1, K.OA.2, SMP1, SMP2, SMP6**

# Compare Body Heights to Objects

## Measurement and Data

Ask children to estimate where their heights will fall compared to taller objects, or how high the heights of shorter objects will be compared to them. For example, a child might be as tall as the windows on a door, or a book may come up to a child's knee.

CCSS K.MD.1, K.MD.2

## Operations and Algebraic Thinking **Number Stories**

Remember to use your children's names when telling number stories.

*Andrew, Martha, John, Lucie, and Nadia ate oranges for lunch. How many girls ate oranges for lunch?* (3 girls) *How many boys ate oranges for lunch?* (2 boys) *How many children all together ate oranges for lunch?* (5 children)

*Carlos wanted some iced tea. He put 3 ice cubes in a glass and poured hot tea over them. How many ice cubes does he have now?* (0 ice cubes) *What happened to them?* (They melted.) *What happened to the tea?* (It cooled off.)



CCSS **K.CC.4, K.CC.5, K.OA.1, K.OA.2, SMP1, SMP2**

# Count and Tap

## Counting and Cardinality

Have children form a circle. Walk around the circle, slowly counting out loud from 1 or another number. At several points, tap a child and have that child say the next number.

CCSS K.CC.1, K.CC.2, SMP6

# In the Yard

## Operations and Algebraic Thinking Number Stories

Children can use their fingers to show the number of animals and acorns in these problems.

*José saw 1 squirrel, 2 chipmunks, and a sandbox in the backyard. How many animals did he see?* (3 animals)

*There were 5 acorns. The squirrel took 4 of them. How many did he leave behind?* (1 acorn)



CCSS K.OA.1, K.OA.2, SMP1, SMP2, SMP4

## Clap and Count Clues

### Counting and Cardinality

Direct one child to clap or stomp to a secret number you show him or her—one clap or stomp for each number. The rest of the children count silently and then say the number aloud.

CCSS K.CC.1, K.CC.4, K.CC.5

## Comparing Ages

### Operations and Algebraic Thinking **Number Stories**

Display the numbers for the second and third problems below if possible.

*Danielle is older than David. David is older than Sarah. Who is older: Danielle or Sarah?* (Danielle) *How do you know?*

*Sarah is 3 years old. David is 2 years older. How old is David?* (5 years old)

*Danielle is 6 years old. David is 5 years old. How much older is Danielle?* (1 year older)



**CCSS K.CC.3, K.CC.7, K.OA.1, K.OA.2, SMP1**

## Foot-Tap Counting

### Counting and Cardinality

Say: *Count with me to 10* (*to 17, to 21,* and so on). Have children tap their feet one time for each number as they count out loud.

CCSS K.CC.1, K.CC.4

# Ages

## Operations and Algebraic Thinking

Ask children:

*How old are you?*

*How old will you be next year?*

*How old were you last year?*

*How old will you be in 3 years?*

*How did you figure out these answers?*



CCSS K.CC.2, K.OA.1, K.OA.2, SMP1, SMP2

# Finger Math Combinations

## Operations and Algebraic Thinking

Have children hold up the correct number of fingers as you say:

*Show me 4 with one hand.*

*Show me 4 with two hands.*

Next hold up fingers and have children mirror you and count the fingers. Use fingers on one hand or on both hands, making different finger combinations for the same number. Ask children to describe the finger combinations. For example:

**7:** 3 fingers on one hand and 4 fingers on the other, or
2 fingers on one hand and 5 fingers on the other

**3:** 3 fingers on one hand, or
2 fingers on one hand and 1 finger on the other

CCSS **K.CC.5, K.OA.1, K.OA.2, K.OA.3, SMP1, SMP2**

## Say the Next Number

### Counting and Cardinality

Begin counting from 1. Stop counting and point to a child who then says the next number in sequence. Continue counting and stop again. Repeat the process, stopping at different numbers.



CCSS K.CC.1, K.CC.2

## Shapes in the Room

### Geometry

Call on children to name 3 objects in the room that look like circles. Repeat with triangles and rectangles (including squares).

CCSS K.G.1, K.G.2, SMP7

## Nod Your Head

### Counting and Cardinality

Say: *Nod your head and count with me to 7* (or *12, 17, 21,* and so on).

Have children nod one time for each number as they count out loud.



CCSS K.CC.1, K.CC.4

## Let's Count—Forward and Backward

### Counting and Cardinality

Lead children in various forms of counting on. Say: *Let's count to…*

*7, beginning at 1.*

*8, beginning at 2.*

*9, beginning at 3.*

*8, beginning at 5.*

Then say: *Now listen carefully.*

*Let's begin at 8 and count down to 3.*

*Let's begin at 9 and count down to 1.*

CCSS **K.CC.1, K.CC.2**

## Family Math

### Operations and Algebraic Thinking Number Stories

Lamont's baby sister weighed 7 pounds. Kim's baby brother weighed 1 pound less. How much did Kim's baby brother weigh? (6 pounds)

Jeff has 3 sisters. He has 2 brothers. Which does he have more of, sisters or brothers? (Sisters) (Display these numbers if possible.) How many sisters and brothers does he have all together? (5 sisters and brothers)

Does his family have more girl children or more boy children? (The same; 3 boys and 3 girls)



CCSS K.CC.4, K.CC.7, K.OA.2, SMP1

## Show a Card

### Counting and Cardinality

Show a card with a number on it and ask what number comes next. Have children "sky-write" it in the air using large arm motions or "finger-write" it on their own palms. Repeat with new numbers.

CCSS K.CC.2, K.CC.3

# More Count and Tap

## Counting and Cardinality

While the class is waiting in line, have children count out loud slowly. At several points, tap a child and have that child give the next number.

 CCSS K.CC.1, K.CC.2

# Melting to Zero

## Operations and Algebraic Thinking **Number Stories**

Janine had 2 ice pops on a hot summer day. She left them outside and forgot about them when her friend came over to play. Both ice pops melted. How many ice pops did she have left when she went back to get them? (She had zero, or none.)

Nick left his glass on the porch on a hot sunny day. It had 5 ice cubes in it. He came back later to get the glass, and all the ice cubes had melted. How many ice cubes were left? (There were zero, or none.) How many ice cubes had melted? (Five ice cubes had melted.)

CCSS **K.OA.1, K.OA.2, SMP1**

## Block Towers

### Measurement and Data

Roberto and Tami were playing with blocks in the block corner. Roberto built a tower with 4 blocks. Tami built a tower with 6 blocks. They used blocks that were the same size and shape and stacked them the same way.

Whose tower was taller? How much taller was it? (Tami's tower; 2 blocks taller)



CCSS K.OA.2, K.MD.1, K.MD.2, SMP1

## Guessing Numbers Drawn on Backs

### Counting and Cardinality

Draw a number with your finger on a child's back. Have the child guess the number.

Then have children take turns finger-drawing numbers on a partner's back and having the partner guess the number. If some children need added practice in "feeling" numbers, have their partners say the numbers as they draw them.

CCSS K.CC.3

# Counting On

## Counting and Cardinality

Say: *Let's count to…*

*12, starting at 5.*

*15, starting at 6.*

*21, starting at 9.*

*9, starting at 3.*

*21, starting at 0.*

 CCSS K.CC.1, K.CC.2, SMP6

# The Number Before

## Counting and Cardinality

Ask: *What is the number before 12?* (11)

*What is the number before 11?* (10)

*What is the number before 10?* (9)

Say: *Let's count backward together from 12: 12, 11, 10, . . .*

CCSS **K.CC.1**

# Counting and Clapping

## Counting and Cardinality

Have children clap and count with you to 15 (or 20, 25, and so on). You may also want to use a funny voice that you or children suggest, such as a baby voice, a giant voice, a big-bad-wolf voice, and so on.

# Numbers Before and After

## Counting and Cardinality

Ask: *When you count, which number follows 3? 15? 12? 8?*

*When you count, which number comes after 7? 11? 9? 10?*

*When you count, which number comes before 3? 5? 8? 6?*

CCSS K.CC.1, K.CC.2, SMP1

# About How Many Children?

## Counting and Cardinality

Pose the following questions to children:

*Without counting, can you tell me about how many girls are here today? More than 10? Fewer than 10?*

*About how many boys are here today? Let's count and check!*

*Is the number of girls greater than, less than, or equal to the number of boys?* (You can have the girls line up next to the boys to compare the groups.)

*How many more boys* (or *girls*) *are here today?* (Model how to count the "extra" children to find out how many more.)



CCSS K.CC.4, K.CC.5, K.CC.6, K.CC.7, SMP4

# Which Is Greater? Which Is Less?

## Counting and Cardinality

Ask: *Which number is greater: 3 or 5? 7 or 4? 1 or 2? 2 or 10?* (Display each pair of numbers if possible.)

Have children explain their answers. (For example: 5 is greater than 3 because when you count, you get to 5 after you've already counted 3.)

Ask: *Which number is less: 3 or 5? 6 or 8? 9 or 10? How do you know?*

CCSS **K.CC.7, SMP3, SMP6**

## Give Numbers Greater Than and Less Than

### Counting and Cardinality

Pose the following prompts. Display the numbers if possible.

*Say a number that is greater than 3. How do you know it is greater than 3?* (Because it comes after 3 when I count; because it's to the right on the Growing Number Line)

*Say another number greater than 3, and another.*

*Say a number that is less than 5.*

*Say another number less than 5, and another.*



CCSS **K.CC.1, K.CC.2, K.CC.7, SMP1, SMP3**

# Number Stories with 1 More

## Operations and Algebraic Thinking Number Stories

Jo has 2 dolls. Cheryl has 1 more doll than Jo has. How many dolls does Cheryl have? (3 dolls)

My tomato plant had 7 tomatoes on it. Then it grew 1 more tomato. Now how many tomatoes are there? (8 tomatoes)

CCSS K.CC.4, K.OA.1, K.OA.2, SMP1

## Lunchtime Geometry

### Geometry

Have the children look for shapes in their lunches (lunch box, sandwich, juice can, fruit, milk carton, and so on).

See if they can change any of the shapes by nibbling (for example, biting a rectangular sandwich into a triangle). Ask them to compare the original shape and the new shape. Ask: *How are they the same? How are they different?*



CCSS K.G.1, K.G.2, K.G.5, SMP6, SMP7

## Standing in Line

### Counting and Cardinality

Call several children to stand in a line at the front of the class. Ask:

*How many children are standing in line? How do you know? Who is standing first? Third? Fourth? 10th?* (and so on).

You may wish to have children count off out loud to determine their positions.

CCSS K.CC.1, K.CC.2, K.CC.4, K.CC.5

# Letters in Order

## Counting and Cardinality

Display the letters *A, B, C, D, E, F, G.*

Ask: *If A is first, what is B?* (Second) *E?* (Fifth) *G?* (Last, or seventh)

*If we go the other way, so that G is first and A is last, what is second?* (F) *What is C?* (Fifth) *B?* (Sixth)

Check to be sure the children know which letter is the "first" one before asking for the second, third, and so on.



CCSS **K.CC.3, K.CC.4, SMP1, SMP2**

# Reading Class Graphs

## Measurement and Data

Have children answer questions about graphs they made as part of lessons in the *Teacher's Lesson Guide*. For example, using the Class Age Graph (Lesson 1-8), ask:

*How many children are 5 years old? How many children are 6 years old?*

*Are more children 5 or 6 years old right now?*

*How many more 5 year-olds are there? How did you figure that out?*

CCSS **K.CC.6, K.CC.7, K.MD.3, SMP4, SMP6**

# Making 10 with Fingers

## Operations and Algebraic Thinking

Show children any number of fingers, from 1 to 10. Ask a volunteer to show you the number of fingers they should add to make 10. Have children count all the fingers to see if they made 10 total. Repeat several times, starting with different numbers of fingers.



CCSS K.CC.4, K.CC.5, K.OA.3, K.OA.4, SMP1, SMP2

## Drawing Numbers with a Foot

### Counting and Cardinality

When children are standing in line, call out different numbers and have the children use a foot to "draw" the numbers on the floor.

CCSS K.CC.3

## Number-Writing on Hands

### Counting and Cardinality

Ask questions about numbers of objects that children see, such as how many chairs are at a table or how many windows are in a room.

Have children "finger-write" the numbers on their palms.



CCSS K.CC.3, K.CC.4, K.CC.5

## Sorting into Groups

### Measurement and Data

Say: *Look at that bookshelf. What different types, or categories, of books do you see? How could we sort them?* (Hardcover/softcover; thick/thin; fiction/nonfiction)

*Look at all of the shoes that we are wearing. How could we sort them?* (Sneakers/loafers; laces/no laces; by color)

CCSS **K.MD.3, SMP1**

# Let's Count On

## Counting and Cardinality

Say: *Let's count to 10 starting with 1.*

*Let's start with 3 and count on to 15.*

*Now let's start with 2 and count on to 12.*



CCSS K.CC.1, K.CC.2

# More Number-Writing on Hands

## Counting and Cardinality

Ask questions about numbers of objects that children see, such as how many tables or chairs there are in the room or how many crayons or markers are in a box.

Have children "finger-write" the numbers on their palms.

CCSS K.CC.3, K.CC.4, K.CC.5

# "Sky-Writing" Shapes

## Geometry

Name some 2-dimensional shapes (triangle, rectangle, square, circle, and so on). Have children "sky-write" each shape as they say its name. Then have them try to identify examples of the shapes in the classroom.



CCSS K.G.1, K.G.2, K.G.5, SMP4

## Part 2

The following activities, along with previous *Minute Math* activities, reinforce the Daily Routines and the lessons through Section 6 in the *Teacher's Lesson Guide*. Remember that any *Minute Math* activity can be used with appropriate changes as often as you would like throughout the year.

# Notes

## What Number Comes After? (10–21)

### Counting and Cardinality

Have children give the number that comes after 7, after 3, after 8, and so on.

This can lead into counting from 10 to 21. Ask: *What comes after 10? After 11? After 12? After 20?* Then count from 10 to 21.

After they practice counting forward from 10 to 21, help the children count backward from 10 to 0.

CCSS **K.CC.1, K.CC.2**

# Greater Than . . .

## Counting and Cardinality

Tell children: *Name any number greater than 10.* Then say: *Give the number that comes next.*

Repeat using different numbers and giving a turn to as many children as possible.



CCSS K.CC.1, K.CC.2

# Find the Next Number—Teens

## Counting and Cardinality

Ask: *When you count, what number comes after 10? After 15? After 17? After 12?*

*When you count, what number follows 13? Follows 16? What number comes after 19? After 14? After 11? After 18?*

# Comparing Shapes

## Geometry

Encourage children to notice and describe properties that are similar between objects, as well as those that differ. For example, ask:

*How is a window like a book? How is it different?*

*How is a penny like the (round) clock? How is it different?*

*How is a book like a table? How is it different?*

*How is a door like a whiteboard? How is it different?*



CCSS **K.G.1, K.G.2, K.G.4, SMP7**

## “Sky-Writing” Numbers

### Counting and Cardinality

Call out different numbers (focus on 0–20) and have the children use large arm motions to “sky-write” the numbers.

CCSS K.CC.3

# Finger Addition

## Operations and Algebraic Thinking

Hold up 3 fingers on one hand and 1 finger on the other. Bring your hands together.

Ask: *How many fingers all together?*

Hold up 2 fingers on one hand and 2 fingers on the other.

Ask: *How many fingers all together?*

Then ask: *Why are they both 4? Can you think of another way to show 4?* (4 fingers on one hand and 0 fingers on the other)



CCSS **K.CC.4, K.CC.5, K.OA.1, K.OA.2, K.OA.3, K.OA.5**

## Guess My Teen Number

### Counting and Cardinality

Say: *I'm thinking of a teen number less than 18.*

Let children take turns guessing what it is. Tell them when they are getting "hotter" (closer to the number) or "colder" (further from the number).

Repeat with other teen numbers. Use a number line as a visual reference if needed.

CCSS K.CC.7, SMP1

# Joining and Giving Away

## Operations and Algebraic Thinking Number Stories

If needed, encourage children to use their fingers to show the numbers of objects in each problem below:

*David has 3 toy cars. Therese has 4 toy cars. How many toy cars do they have all together?* (7 toy cars)

*If you have 2 crayons and you give your neighbor 1 crayon, how many crayons will you have left?* (1 crayon)

*Eduardo had 9 dollars. Then he earned 1 dollar more. How many dollars does Eduardo have now?* (10 dollars)


CCSS **K.OA.1, K.OA.2, SMP1, SMP2**

# Leaving Out Numbers

## Counting and Cardinality

Say: *I will count and leave out one number. Tell me what number I left out.*

*0, 1, 3, 4, 5* (2)

*12, 13, 14, 16* (15)

*21, 22, 24, 25* (23)

CCSS **K.CC.1, K.CC.2, SMP1, SMP6**

## “I Spy a Shape”

### Geometry

Choose an object and say: *I spy a square* (or *circle, triangle, rectangle,* and so on).

Give clues using position words: *It's in the front of the room. It's near the ceiling.*

Have children try to guess your object.

When they are familiar with the game, let children choose the objects and give the clues.



CCSS **K.G.1, K.G.2, SMP6**

# Let's Count—20s and 30s

## Counting and Cardinality

Say: *Let's count to…*

*26, starting at 16.*

*31, starting at 25.*

*35, starting at 21.*

*40, starting at 28.*

# Greater Than and Less Than

## Counting and Cardinality

Say the number 13; display it if possible. Then give children prompts such as:

*Give a number greater than 13. Write the number in the air or in your palm. How do you know it is greater? Count up from 13 to your number.*

*Give a number that is less than 13. Write the number in the air or in your palm. How do you know it is less? Count up from your number to 13.*



CCSS K.CC.2, K.CC.7, SMP3, SMP6

## Counting Down

### Counting and Cardinality

Say: *Let's count down from 10 to 5; 14 to 9;* and so on.

Encourage children to refer to the Growing Number Line or Number-Grid Poster if they are available.

CCSS **K.CC.1, K.CC.2, SMP5**

## Counting Backward—What Is Missing?

### Counting and Cardinality

Say: *I'm going to count backward from 7: 7, 6, 5, 3, 2, 1. Did I leave out any numbers?* (Yes; 4)

*How would you count backward from 7?* (Ask individual students.)

*Let's do it together: 7, 6, 5, 4, 3, 2, 1.*



CCSS K.CC.1, K.CC.2, SMP6

## "I Spy 3-Dimensional Shapes"

### Geometry

After you have introduced 3-dimensional shapes, play *I Spy* with those shapes. Also include other attributes in your clues.
For example:

*I spy a red sphere.*

*I spy a small green cube.*

*I spy a cylinder with black letters and orange pictures.*

CCSS **K.G.1, K.G.2, SMP7**

# Taller, Farther, Longer

## Operations and Algebraic Thinking **Number Stories**

I live in a 6-floor building. My friend lives across the street in an 8-floor building. Whose building is taller? (My friend's building) How much taller is it? (2 floors taller)

Chan ran 3 laps around the track. Josh ran 2 laps. Who ran farther? (Chan) How much farther? (1 lap farther)

Adam and Ariel have new sleds. Adam's sled is 3 feet long, and Ariel's is 4 feet long. Whose sled is longer? (Ariel's sled) How much longer is it? (1 foot longer)



CCSS K.CC.7, K.OA.2, K.MD.2, SMP1

# Which Is Heavier?

## Measurement and Data

Ask: *Which do you think is heavier (weighs more): a mouse or an elephant? A bike or a car? An apartment building or a doghouse?*

*Which do you think is lighter (weighs less): a feather or a brick? A party balloon or a basketball? A tree or a tulip?*

CCSS **K.MD.1, K.MD.2**

## Name a Number

### Counting and Cardinality

Name any number less than 20. Then have children give the number that comes next. Do this with several numbers.

Name another number less than 20. Then have children give the number that comes *before* that number. Do this with several numbers.



CCSS **K.CC.2, K.CC.4, SMP6**

## Counting by 10s

### Counting and Cardinality

Say: *Let's count the fingers in our room by 10s.*

Have each child hold up the fingers of both hands and put them down as they are touched and counted: *10, 20, 30, . . .*

If you have not already done so, mark the 10s on your Class Number Line for easy counting by 10s.

CCSS K.CC.1

# Birthday Party

## Operations and Algebraic Thinking Number Stories

Remember to use your children's names in number stories such as the following:

*Roberto had a birthday party for his sixth birthday. His classmates Kim, Ted, and Tomi were there. His mother and grandfather were also at the party. Including Roberto, how many people were at the party?* (6 people) *How many children were at the party?* (4 children)

*Each child had a birthday hat. How many birthday hats were there?* (4 birthday hats)



CCSS K.CC.4, K.CC.5, K.OA.1, K.OA.2, SMP1

# Say the Next Numbers—1 to 50

## Counting and Cardinality

Begin counting from different numbers. Stop counting, point to a child, and ask him or her to say the next 3 or 4 numbers in sequence. Stop that child (with a hand signal or stop sign) and point to another child who should continue counting. Keep counting and stopping until you reach 50.

Repeat the process, stopping at different numbers along the way.

CCSS K.CC.1, K.CC.2

# Fall Leaves

## Operations and Algebraic Thinking

**Number Stories**

Maria and Jack found colored leaves under the trees in the park. Maria took home 8 leaves. Jack did not take home as many as Maria. How many leaves did Maria take home? (8 leaves) How many leaves did Jack take home? (We don't know—just that it was less than 8.)



CCSS K.OA.1, K.OA.2, SMP1

## More Let's Count

### Counting and Cardinality

Say: *Let's count to…*

*22, starting at 15.*

*33, starting at 27.*

*42, starting at 35.*

*50, starting at 43.*

*Now let's count to 50 starting at 0.*

# Here's a Riddle

## Counting and Cardinality

Say: *I'm thinking of a number. It's greater than 1. It's greater than 4. It's less than 6. What number is it?* (5)

Create similar riddles for other numbers. Have children create their own number riddles.



CCSS K.CC.7, SMP1

## Give Any 3 Numbers

### Counting and Cardinality

Encourage children to use a number line for the following prompts if needed.

Say: *Tell me any 3 numbers that are less than 12.*

*Tell me any 3 numbers that are greater than 12.*

*Name all of the numbers between 30 and 37; between 14 and 18; between 45 and 49;* and so on.

## First or Last?

### Counting and Cardinality

Sketch a row of pictures on the board (or on cardstock for repeated use).

Ask: *Which figure is at the beginning of the row? At the end?* Then have the children tell which picture is first, third, sixth, and so on.



CCSS K.CC.4, K.CC.5, SMP6

# Finger Math

## Operations and Algebraic Thinking

Hold up some fingers on one hand. Ask: *How many fingers am I holding up?*

Hold up some fingers on the other hand. Ask: *How many fingers do I have up on this hand?*

Bring the hands close together. Ask: *How many fingers are there all together?*

Say: *Show me that number with your fingers in a different way.*

Repeat using other numbers and finger combinations.

**CCSS K.CC.4, K.CC.5, K.OA.1, K.OA.3, SMP1, SMP2**

## "I Spy Shapes in Different Positions"

### Geometry

Play *I Spy* with shapes and position words. For example, give the following clues:

*I spy a circle above the table.*

*I spy a square below the calendar.*

*I spy a rectangle next to the chair.*

Clues can also include position words such as *beside, in front of,* and *behind.*



CCSS K.G.1, K.G.2, SMP6, SMP7

# At the Zoo

## Operations and Algebraic Thinking Number Stories

At the zoo, Sandra saw one snake that was 4 feet long, a second snake that was 2 feet long, and a third snake that was as long as the first and second snake together. How long was the third snake? (6 feet long) Which snake was the longest? (third snake) How do you know? Which snake was the shortest? (second snake) How do you know?

Sherry had 6 pennies to spend at the gift shop. She bought a tiger eraser for 2 pennies. Does Sherry have less money now? (Yes.) How much money does Sherry have now? (4 pennies)

CCSS K.OA.1, K.OA.2, K.MD.2, SMP1, SMP3

# Numbers Before, After, and Between

## Counting and Cardinality

Say: *Give the two numbers that come after 23; after 35; after 49; before 16;* and so on.

*Give all the numbers from 33 to 39; 12 to 18; 7 to 3; 12 to 1.*



CCSS K.CC.1, K.CC.2

## More Number-Writing on Hands

### Counting and Cardinality

Ask questions about numbers of objects, such as how many shoes are in a pair, how many fingers are on one hand, or how many toes are on two feet.

Have children "finger-write" the numbers on their palms.

CCSS K.CC.3, K.CC.4, K.CC.5

## "I'm Thinking of a Shape" (2-Dimensional)

### Geometry

Think of a 2-dimensional shape and describe its characteristics. Then have children try to guess the shape. For example: *I'm thinking of a shape that has 3 straight sides. What is its name?* (Triangle)

When children are familiar with the game, they can think of the shapes and give the clues.



CCSS K.G.2, K.G.4, SMP7

# Tell a Number Story

## Operations and Algebraic Thinking

Number Stories

Invite a child to tell a 6 story (a story that has 6 as the answer). For example: *There are 2 kids at the park, and 4 more join them. How many kids are there all together?*

Then invite another child to tell a different 6 story. Compare the stories.

Repeat with other stories, such as a 1 story or a 5 story.

CCSS K.OA.1, K.OA.2, K.OA.3, SMP1

## Using "Equals"

### Operations and Algebraic Thinking

Display a row of numbers: 1, 2, 3, 4, 5, 6, 7, 8, 9, 10. Point to particular numbers and say: *Give the number that comes 2 numbers after this one; 3 numbers after this one; 1 number before this one;* and so on.

As children respond, paraphrase what is happening by using the terms *makes, is,* and *is equal to* interchangeably. For example: *6 and 2 more* makes *8; 1 less than 5* is *4; 3 and 3 more* is equal to *6.*



CCSS K.CC.4, K.OA.1, K.OA.2, SMP2, SMP6

# What Number Comes Between?

## Counting and Cardinality

Ask: *What number comes between 2 and 4? 11 and 13? 14 and 16? 18 and 20?*

*What numbers come between 13 and 17? 21 and 24?*

## Erasing Slates

### Operations and Algebraic Thinking **Number Stories**

Remember to use your children's names in number stories such as the following below and to vary the numbers to meet your class's needs. Also remind children to use units (children) with the numbers in their responses.

*Eddy's teacher asked 2 boys and 3 girls to erase the slates. How many children were asked to erase the slates?* (5 children) *How did you figure it out?*



CCSS K.CC.5, K.OA.1, K.OA.2, SMP1, SMP6

# Give 3 Numbers Before and After

## Counting and Cardinality

Say: *Give the 3 numbers that come after 12; after 26; after 30; after 47; after 52;* and so on.

*Counting backward, give the 3 numbers that come before 6; before 10; before 16;* and so on.

CCSS K.CC.1, K.CC.2

# More What Number Comes Between?

## Counting and Cardinality

Ask: *What number comes between 12 and 14? 19 and 21? 34 and 36? 48 and 50?*

*What numbers come between 18 and 21? 24 and 27? 50 and 56?*



CCSS K.CC.1, K.CC.2, SMP1

# Reading Class Graphs

## Measurement and Data

Ask children to answer questions about graphs made previously in the classroom.

For example, using the Pattern-Block Graph (Lesson 3-1), ask:

*How many triangles are there? How do you know?*

*How many more squares are there than trapezoids? How do you know?*

*Which shape was most common (had the greatest number)? Which shape was least common (had the fewest number)? How do you know?*

CCSS **K.CC.5, K.CC.6, K.MD.3, K.G.2, SMP2, SMP4, SMP6**

# Fish

## Operations and Algebraic Thinking **Number Stories**

Donna's aquarium has 3 angelfish and 2 catfish. Can we tell how many fish are in her aquarium? (Yes.) How many? (5 fish)

Her friend just gave her 2 goldfish. Now how many fish are in her aquarium? (7 fish)

How many more fish would she need to get to 10 fish in her aquarium? (3 more fish)



CCSS K.OA.1, K.OA.2, K.OA.4, SMP1

# More Standing in Line

## Counting and Cardinality

This is a good activity to do when you are waiting in line.

Have the class count how many children are standing in line. Then ask: *Who is standing first? Second? Last? Fifth? 10th? 15th?* You may wish to have children count off out loud to figure out their positions.

CCSS **K.CC.1, K.CC.2, K.CC.4, K.CC.5**

# Numbers That Come Before

## Counting and Cardinality

Encourage children to look at the Growing Number Line, or to picture it in their heads, for prompts such as the following:

*When you count, what number comes before 13? 15? 18? 16? 20? 11? 17? 10?*



CCSS K.CC.1, K.CC.2

## Lunchtime

### Operations and Algebraic Thinking

Number Stories

Ask children to share and compare their strategies for solving the following problems:

*It was lunchtime. Tom had 7 carrot sticks. Carla had 5. How many more did Tom have?* (2 more carrot sticks)

*Rex ate 10 blueberries. Then he ate 2 more. How many blueberries did he eat in all?* (12 blueberries)

*Fred had several orange slices. After sharing 3 of his orange slices, he had 2 left. How many orange slices did he start with?* (5 orange slices)

CCSS **K.OA.1, K.OA.2, SMP1, SMP6**

# More Name a Number

## Counting and Cardinality

Tell children to name any number less than 30, then give the number that comes next. (Do this with several numbers.)

Then tell them to name any number between 30 and 40, then give the number that comes next. (Do this with several numbers.)



CCSS K.CC.1, K.CC.2

# Give Numbers After and Between

## Counting and Cardinality

Say: *Give the 2 numbers, in order, that come after 13; after 25; after 50; after 67;* and so on.

*Give all the numbers from 13 to 19; 42 to 48; 17 to 13; 19 to 10.*

CCSS K.CC.1, K.CC.2, SMP6

# Tell Me about 2-Dimensional Shapes

## Geometry

Say: *We are going to tell each other about 2-dimensional (flat) shapes.*

*I'll tell you about a square. It has 4 vertices and 4 sides of equal length.*

*Tell me about a triangle.*

*Tell me about a rectangle.*

*Tell me about a hexagon.*

*Tell me about a circle.*



CCSS **K.G.2, K.G.4, SMP6, SMP7**

## More Counting by 10s

### Counting and Cardinality

Say: *Let's count by 10s to 100.*

# Numbers After and Before

## Counting and Cardinality

Have children do this activity with or without a number line. Encourage them to picture a number line in their heads if needed.

Say: *Give the 3 numbers that come in order after 14; after 37; after 51; after 65;* and so on.

*Counting backward, give the 3 numbers that come before 15; before 18; before 10; before 20;* and so on.



CCSS K.CC.1, K.CC.2, SMP5, SMP6

# More Numbers After and Before

## Counting and Cardinality

Have children do this activity with or without a number line. Encourage them to picture a number line in their heads if needed.

Say: *Give the 3 numbers that come in order after 38; after 49; after 62;* and so on.

*Counting backward, give the 3 numbers that come before 14; before 11; before 20;* and so on.

CCSS K.CC.1, K.CC.2, SMP5, SMP6

# Kittens

## Operations and Algebraic Thinking **Number Stories**

The Garcias have a family of 7 cats: a mother, a father, and their kittens. How many kittens are there? (5 kittens)

Brenda Garcia gave away 3 of the kittens to happy homes. How many kittens did she keep? (2 kittens)



CCSS **K.OA.1, K.OA.2, SMP1**

## More Finger Addition

### Operations and Algebraic Thinking

Say: *Everyone hold up 3 fingers on one hand.*

*Hold up 2 fingers on your other hand.*

*Bring your hands together.*

*How many fingers are you holding up?* (5)

*Use your fingers to show 5 a different way.* (0 and 5, 1 and 4, 2 and 3, 4 and 1, 5 and 0)

CCSS **K.CC.5, K.OA.1, K.OA.2, K.OA.3, K.OA.5, SMP1, SMP2**

## Give Numbers in Order

### Counting and Cardinality

Say: *Give the numbers in order from 61 to 67; 52 to 58; 63 to 69; 55 to 58; 18 to 13; 19 to 11;* and so on.

*Now let's all count together from 1 to 70. Let's clap on every 10th number (10, 20, 30, and so on).*



CCSS **K.CC.1, K.CC.2, SMP6**

# Give the Next Number

## Counting and Cardinality

Call on children to quickly answer the following. Encourage them to reference a number line or number grid if needed.

Ask: *When you count, what number comes after 75? After 82? After 58? After 33? After 50?*

*When you count, what number follows 64? 72? 81? 47? 13?*

CCSS K.CC.1, K.CC.2, SMP5

# Counting by 10s and 5s

## Counting and Cardinality

Use children's hands to help them practice counting by 10s. Say: *Both hands together have 10 fingers. Let's count by 10s to see how many fingers are in the room.* Have children hold up both hands and put down their two hands as they are touched and counted. Lead the class: *10, 20, 30, 40, . . .*

If desired, also help them count by 5s. Say: *Each hand has 5 fingers. Let's count by 5s to see how many fingers are in the room.* Have the children hold up both hands and put each one down as it is touched and counted. Lead the class: *5, 10, 15, 20, 25, 30, 35, 40, . . .*

If you have not already done so, mark the 10s and 5s on your Growing Number Line for skip counting.



CCSS **K.CC.1, K.CC.5**

## Name Any 3 Numbers

### Counting and Cardinality

Children may find it helpful to use a number line for the following prompts:

*Name any 3 numbers less than 55.*

*Name any 3 numbers greater than 55.*

*Name all the numbers between 13 and 17; 23 and 27; 43 and 47;* and so on.

CCSS K.CC.1, K.CC.2, K.CC.3, K.CC.7, SMP5

# More Counting by 10s and 5s

## Counting and Cardinality

Emphasize the chant while leading the class in skip counting:

*Let's count to 50 (or more) by 10s: 10, 20, 30, 40, 50.*

*Let's count to 50 (or more) by 5s: 5, 10, 15, 20, . . .*

# Give 3 Numbers That Follow

## Counting and Cardinality

Say: *Count the 3 numbers that follow 76; 44; 88;* and so on.

*Counting backward, give the 3 numbers that come before 17; 29; 53;* and so on.

CCSS **K.CC.1, K.CC.2, SMP6**

# Counting Between Numbers

## Counting and Cardinality

Say: *Count from 14 to 18; 27 to 32; 61 to 67; 88 to 90; 40 to 46; 47 to 51;* and so on.

*Count down from 38 to 34; 20 to 16; 14 to 10.*



CCSS K.CC.1, K.CC.2, SMP6

## On the Playground

### Operations and Algebraic Thinking **Number Stories**

You may wish to act out the following stories as children try to solve them, or afterward:

*Three girls were on the playground, playing with some boys. There was 1 more boy than there were girls. How many boys were there?* (4 boys) *How many children were there all together?* (7 children)

*The 7 children were playing. Then 1 child left to go inside. Then 2 more children went inside. How many children went inside?* (3 children)

*How many children were left outside?* (4 children)

CCSS **K.CC.4, K.OA.1, K.OA.2, SMP1**

## Drawing Shapes

### Geometry

While the class is moving to the carpet or getting in line, have a child draw a triangle on the board, on chart paper, or on a slate. Challenge other children to draw triangles that differ in orientation or size. Ask: *Why are these all triangles?*



CCSS **K.G.2, K.G.4, K.G.5, SMP2, SMP4, SMP7**

# How Many More? How Many Left?

## Operations and Algebraic Thinking Number Stories

One day, 5 crates of chocolate milk and 3 crates of regular milk were delivered to the school lunchroom. Did they bring more crates of chocolate milk or more crates of regular milk? (Chocolate milk) How many more? (2 more crates)

Today, 4 children will get turns to play on the computer. Three of those children have already had their turns. How many children still need a turn? (1 child)

# Counting Up and Back

## Counting and Cardinality

Say: *Count the numbers from 34 to 38; 57 to 62; 86 to 90; 58 to 63;* and so on.

*Count back from 16 to 9; 46 to 38; 81 to 77.*



CCSS K.CC.1, K.CC.2, SMP6

# How Many More?

## Operations and Algebraic Thinking Number Stories

Have children count on using their fingers for these stories if they wish.

*A shirt has 6 buttons. It needs 9 buttons. How many more buttons does it need?* (3 more buttons)

*There are 8 hot dogs in a package. How many more hot dogs does it take to make 10?* (2 more hot dogs)

CCSS K.OA.1, K.OA.2, SMP1, SMP2

# Pets

## Operations and Algebraic Thinking

Number Stories

Children can use their fingers or other objects to show the numbers of animals in each of these problems.

*Toshi has 2 cats, 3 rabbits, and 1 bird. How many animals does she have all together?* (6 animals)

*Toshi gave the 3 rabbits to her class as classroom pets. How many animals does she have left?* (3 animals)

*The class has 3 rabbits. How many more rabbits do they need if they want to have 5 rabbits all together?* (2 more rabbits)



CCSS **K.OA.1, K.OA.2, SMP1, SMP2**

## How Many Pieces?

### Operations and Algebraic Thinking Number Stories

David had 12 pieces of chalk. His brother Abe had 10. Who had more pieces of chalk, David or Abe? (David) How many more pieces did he have? (2 more pieces)

David and Abe were given 6 balloons to share equally. How many should each child have? (3 balloons) How did you figure it out?

CCSS K.OA.1, K.OA.2, SMP1

# Tell Any 3 Numbers

## Counting and Cardinality

Children may find it helpful to use a number line for these prompts:

*Tell any 3 numbers less than 80.*

*Tell any 3 numbers greater than 80.*

*Let's start at 50 and count to 80 by 1s.*



CCSS K.CC.1, K.CC.2, K.CC.3, K.CC.7, SMP5

# Sharing Pennies

## Operations and Algebraic Thinking

Number Stories

Malik and Robby are best friends. One day Malik brought 4 pennies to school and Robby brought 5. Robby wanted his friend to have as many pennies as he had, so he gave Malik 1 of his pennies. Did they both have the same number of pennies then? (No.) Why not? (Have two children act this out.)

CCSS K.CC.6, K.OA.1, K.OA.2, SMP1, SMP2, SMP6

# Find Numbers Before, After, and Between

## Counting and Cardinality

Say: *Give the numbers before and after 55; 81; 20; 63; 19;* and so on.

Then ask: *What numbers come between 56 and 64? Between 73 and 68? Between 37 and 42? Between 79 and 82?*

 CCSS K.CC.1, K.CC.2, SMP6

## How Many Pencils?

### Operations and Algebraic Thinking **Number Stories**

Children can use their fingers to show the numbers of pencils in these problems:

*Nate sharpened 5 pencils. Luis sharpened 3 pencils. Which boy sharpened more pencils?* (Nate) *How many more?* (2 more pencils)

*Johanna can find only 2 of her 5 pencils. How many pencils are missing?* (3 pencils)

*Lori has 2 blue pencils and 5 yellow pencils. How many pencils does Lori have?* (7 pencils) *How many more yellow pencils than blue ones does Lori have?* (3 more yellow pencils)

CCSS **K.CC.6, K.OA.1, K.OA.2, SMP1, SMP2**

# Give Numbers After and Before

## Counting and Cardinality

Ask: *When you count, what is the number that comes after 65? After 89? After 19?*

Then say: *Name any number that comes before 31; before 74; before 51; before 60.*



CCSS **K.CC.1, K.CC.2, SMP6**

## “I’m Thinking of a Shape”

### Geometry

Say: *I’m thinking of a shape.* Give one clue at a time until a child guesses the answer.

*It has four sides.*

*All four sides are straight.*

*All four sides are the same length.*

*What is it?* (A square)

Repeat with other shapes.

CCSS **K.G.2, K.G.4, SMP6, SMP7**

# Walking to School

## Operations and Algebraic Thinking Number Stories

Martin walks 4 blocks to school and 4 blocks home each day. How many blocks does he walk to and from school in one day? (8 blocks)

On his way to school, Martin noticed that the thermometer outside the school read 30 degrees. On his way home, it read 40 degrees. How many degrees did the temperature rise? (10 degrees)



CCSS K.OA.1, K.OA.2, SMP1

# Say Numbers Before, After, and Between

## Counting and Cardinality

Say: *Name the numbers before and after 23; 41; 60; 20; 12;* and so on.

Then ask: *What numbers come between 28 and 32? Between 47 and 51? Between 82 and 87? Between 89 and 91?*

## More Leaving Out Numbers

### Counting and Cardinality

Say: *I will say some numbers and leave out one number. Tell me which number I left out.*

*51, 52, 54, 55* (53)

*36, 38, 39, 40* (37)

*I will give a string of numbers and mix up two of the numbers. Tell me which numbers I mixed up.*

*68, 69, 70, 72, 71* (72 and 71)

*21, 20, 22, 23, 24* (21 and 20)

 CCSS **K.CC.1, K.CC.2, SMP6**

# Name Any 3 Numbers

## Counting and Cardinality

Say: *Name any 3 numbers that are less than 85. Name any 3 numbers that are greater than 85.*

*Let's count from 55 to 85.*

Children may find it helpful to use a number line.

CCSS **K.CC.1, K.CC.2, K.CC.3, K.CC.7, SMP5, SMP6**

# Find Numbers Between

## Counting and Cardinality

Say: *Name 5 different numbers that are greater than 38 and less than 44.*

*Name 5 different numbers that are greater than 64 and less than 70.*



CCSS K.CC.1, K.CC.2, SMP6

# How Much Money?

## Operations and Algebraic Thinking **Number Stories**

A soccer ball costs $9. It costs $1 more than a football. How much does the football cost? ($8)

Joaquin was given $5 for his birthday. He found that the soccer ball was on sale for $6. How much more money does he need to buy the soccer ball? ($1 more)

Joaquin's uncle gave him $1. Does he have enough money to buy a soccer ball now? (Yes.) Does he have any extra money? (No.)

# Equivalencies and Straws

## Operations and Algebraic Thinking **Number Stories**

Pose the following problem: *Jamal has 6 straws. What are the different ways he could hold them using one or both hands?* (6 in one hand and 0 in the other; 5 and 1; 4 and 2; 3 and 3; 2 and 4; 1 and 5; and 0 and 6)

Repeat with other numbers of straws.



CCSS **K.OA.1, K.OA.2, K.OA.3, SMP1**

## Counting On Using Fingers

### Operations and Algebraic Thinking

Hold up 2 fingers of one hand. Ask: *How many more fingers do I need to make 5?* (3 more fingers)

Repeat with different starting numbers (0–5).

# The Flea Market

## Operations and Algebraic Thinking Number Stories

Room 14 was having a Flea Market. Most of the children brought 1 or 2 old toys to sell.

Emmi brought some cars. They sold for 10 cents each. Saul bought 2 of them. How much did he pay? (20 cents)

Tami sold her blocks and sticker books for 10 cents each. Jean bought 1 block and 2 sticker books. How much did she pay? (30 cents)

Dan sold his tennis balls for 10 cents each. Emmi bought 4 tennis balls. How much did she pay? (40 cents)



CCSS K.CC.1, K.OA.1, K.OA.2, SMP1

# Name Numbers in Order

## Counting and Cardinality

Say: *Name the numbers in order from 34 to 38; 57 to 62; 14 to 9; 95 to 100; 4 to 0; 58 to 63;* and so on.

CCSS K.CC.1, K.CC.2, SMP6

# Can I Buy It?

## Operations and Algebraic Thinking **Number Stories**

Remember to use your children's names in number stories like the following:

*Gina and Paul were counting their money. Gina has 7 cents. She wants to buy a 10-cent ball. How many more cents does she need?* (3 more cents)

*Paul has 5 cents. He would like to buy a magnet that costs 10 cents. How much more money does he need?* (5 more cents)



CCSS **K.OA.1, K.OA.2, K.OA.4, SMP1**

## Think of Any 3 Numbers

### Counting and Cardinality

Say: *Let's count by 10s to 100.*

*Think of any 3 numbers less than 100. "Sky-write" your numbers.*

*Think of any 3 numbers greater than 100. "Sky-write" your numbers.*

CCSS K.CC.1, K.CC.2, K.CC.3, K.CC.7

## More Equivalencies and Straws

### Operations and Algebraic Thinking **Number Stories**

Say: *Malia has 4 straws. What are the different ways she could hold them using one or both hands?* (4 in one hand and 0 in the other, 3 and 1, and so on)

Repeat with other numbers of straws.



CCSS K.OA.1, K.OA.2, K.OA.3, SMP1

# How Long? How Tall?

## Operations and Algebraic Thinking Number Stories

Hannah's hair is 15 inches long. It is 1 inch longer than Tammy's hair. How long is Tammy's hair? (14 inches long) How do you know?

Maria's desk fits under the windowsill in her room. The windowsill is 34 inches high (above the floor). How tall is her desk? (Less than 34 inches tall) How do you know?

CCSS K.OA.1, K.OA.2, K.MD.2, SMP1

# Water Capacity

## Measurement and Data **Number Stories**

Karlis filled a cup with water and poured it into a pot. The pot was only partially full. Which holds more water, the cup or the pot? (The pot) How do you know?

What would happen if he filled that pot and poured it into that cup? (The cup would overflow.) Why do you think that?



CCSS K.MD.2, SMP6

## "I Spy 2-Dimensional Shapes"

### Geometry

Play *I Spy* with shape clues. Also include other attributes. For example:

*I spy a brown square.*

*I spy a large blue rectangle.*

*I spy a sphere with black and white hexagons on it.*

CCSS K.G.1, K.G.2, SMP6, SMP7

# What Comes Next?

## Counting and Cardinality

Ask: *What number comes next after: 9? 16? 23? 37? 44? 52? 65? 78? 81? 89? 94?*

Use different numbers to help children learn the number sequence between 1 and 100. The transition to the decade numbers (10, 20, 30, and so on) is often the trickiest for children, so provide extra practice with those.



CCSS K.CC.1, K.CC.2, SMP6

## Name 5 Numbers

### Counting and Cardinality

Say: *Name 5 different numbers that are greater than 27 and less than 33.*

*Name 5 different numbers that are greater than 52 and less than 58.*

CCSS K.CC.1, K.CC.2, K.CC.7

# Ferris Wheel

## Operations and Algebraic Thinking **Number Stories**

Children can use their fingers to show the number of people for these number stories.

*There were 8 people on a Ferris wheel. After the first ride, several people got off, but 3 people stayed on for another ride. How many people got off?* (5 people) *How did you figure it out?*

*On the next ride, some more people got on. How many people were on the Ferris wheel then?* (We don't know.) *Why don't we know? What other information do we need?* (We need to know how many more people got on for the next ride.)



**CCSS K.OA.1, K.OA.2, SMP1, SMP2**

## How Many More Do I Need?

### Operations and Algebraic Thinking **Number Stories**

Children can keep track with their fingers if needed for these stories:

*I have 4 marbles. I want 10 marbles. How many more marbles do I need?* (6 more marbles) *How did you figure it out?*

*I took out 10 books from the library. I returned 8 of them. How many more books do I still need to return?* (2 more books) *How did you figure it out?*

CCSS **K.OA.1, K.OA.2, K.OA.4, SMP1, SMP2**

# Toy Combinations

## Operations and Algebraic Thinking

Number Stories

Zach has a collection of stuffed dogs and bears. There are 4 animals total. What are the different combinations of stuffed animals that he might have? (1 dog and 3 bears; 3 dogs and 1 bear; 2 dogs and 2 bears)

Jill has a collection of toy trucks and race cars. There are 5 vehicles total. What are the different combinations of vehicles that she might have? (1 truck and 4 race cars; 4 trucks and 1 race car; 2 trucks and 3 race cars; 3 trucks and 2 race cars)



CCSS K.OA.1, K.OA.2, K.OA.3, SMP1

# Greater and Less

## Counting and Cardinality

Children may want to refer to a number line for these activities.

Say: *Tell me a number greater than 50.*

*Tell me another number greater than 50 (and another,* and so on).

*Give two numbers less than 45. Give another number less than 45.*

## Teen Finger Math

### Number and Operations in Base Ten

Hold up 10 fingers.

Have a partner hold up fingers next to yours to make 11; then 14; then 17.

Ask: *How do you know how many to hold up?*

 CCSS **K.NBT.1, SMP2**

# Making Teen Numbers

## Number and Operations in Base Ten

Pose the following prompts to help children think about the numbers 10 through 19.

*10 and how many more make 14?* (4)

*18 is 10 and how many more?* (8)

*10 is 10 and how many more?* (0)

*10 and how many more make 17?* (7)

*10 and how many more make 12?* (2)

*10 and how many more make 19?* (9)

*15 is 10 and how many more?* (5)

*11 is 10 and how many more?* (1)

*16 is 10 and how many more?* (6)

CCSS K.NBT.1

## How Could We Sort Our Class?

### Measurement and Data

Ask: *What categories could we use to sort children in our class?* (Girls/boys; dark/light hair; color of shirts; type of shoes; length of sleeves; and so on)

Together, sort the class using one of the categories. Then ask: *How many children are in each category? Which category has the most children? Which one has the fewest children?*



**CCSS K.CC.4, K.CC.5, K.CC.6, K.MD.3, SMP8**

# What Comes After and Before?

## Counting and Cardinality

Have children do this activity with or without a number line.

Say: *Count the 3 numbers that come after 19; after 25; after 39; after 48;* and so on.

*Count backward. Say the 3 numbers that come before 5; before 18; before 30;* and so on.

CCSS K.CC.1, K.CC.2, SMP5

## More Teen Finger Math

### Number and Operations in Base Ten

Hold up 10 fingers.

Have a partner hold up fingers next to yours to make 13; then 16; then 19.

Ask: *How do you know how many to hold up?*



CCSS K.NBT.1, SMP2

## Ordering a Line

### Counting and Cardinality

While waiting in line, ask:

*Who is first in line?*

*Second from the end?*

*Third from the beginning?*

*Exactly in the middle?*

# More Give 3 Numbers That Follow

## Counting and Cardinality

Say: *Count the 3 numbers that follow 45; 57; 78;* and so on.

*Counting backward, name the 3 numbers that come before 13; 27; 50;* and so on.

*Let's all count to 100 together.*

 CCSS K.CC.1, K.CC.2

## Drawing Shapes in the Air

### Geometry

Have children draw a triangle in the air. Then have them draw a rectangle in the air. Ask several children to describe how the shapes feel different to draw.

CCSS K.G.2, K.G.4, K.G.5, SMP2, SMP4

# Who Is Taller?

## Measurement and Data

Have children compare the heights of two classmates. For example, ask: *Who is taller, Katie or Jamal? How could you find out for sure?*



CCSS K.MD.2, SMP1

## Describing Objects

### Measurement and Data

Tell children: *Look at that table. How many stick-on notes* ***tall*** *do you think it is? How many stick-on notes* ***wide*** *do you think it is? How many stick-on notes* ***long*** *do you think it is?*

You may wish to allow children to measure the table with stick-on notes at another time. Repeat for other familiar objects that are easy for the class to see.

# Snack Combinations

## Operations and Algebraic Thinking

Number Stories

Maggie put out cheese cubes and crackers for her friends. There were 7 snacks total. What are the different combinations of cheese cubes and crackers that she might have had? (1 cheese cube and 6 crackers; 2 cheese cubes and 5 crackers; 3 cheese cubes and 4 crackers; and so on)

Henry put some celery and carrot sticks on his plate. There were 9 vegetable sticks total. What are the different combinations of celery sticks and carrot sticks that he might have had? (1 celery stick and 8 carrot sticks; 2 celery sticks and 7 carrot sticks; 3 celery sticks and 6 carrot sticks; and so on)



CCSS K.OA.1, K.OA.2, K.OA.3, SMP1

## Part 3

The following activities, along with previous *Minute Math* activities, reinforce the Daily Routines and the lessons through Section 9 in the *Teacher's Lesson Guide.* Remember that any *Minute Math* activity can be used with appropriate changes as often as you like throughout the year.

# Notes

## Ten and Some More Line Up

### Number and Operations in Base Ten

Have 10 children line up at the door and count off 1 to 10. Ask: *How many more children need to join the line to make 12?* (2 more children) *How do you know?* Have two more children line up, leaving a space after the 10th child. Have the children count off again.

Repeat during transitions using other teen numbers.

CCSS K.NBT.1

## Break Apart the Line

### Number and Operations in Base Ten

Choose a group of 11–19 children to line up and count off.

Ask where they should break apart the line to show that 10 children and some more children make this number of children in line. (After the 10th child)

Have the first 10 children raise their hands while the remaining children each take one step backward to "break" the line.

Try this at other times with different teen numbers.

# Telling More Number Stories

## Operations and Algebraic Thinking **Number Stories**

Invite a child to tell a 3 story (a story that has 3 as the answer) for the class to solve.

Then invite another child to tell a different 3 story.

Compare the stories.

Repeat with other stories, such as a 0 story or a 5 story.

CCSS **K.OA.1, K.OA.2, K.OA.3, K.OA.5, SMP1**

## Simple Addition

### Operations and Algebraic Thinking

Say: *Solve these using the quickest strategy you are comfortable with:*

*1 + 0*

*3 + 0*

*4 + 0*

*What strategy did you use? What do you notice? What might we call that strategy?*



CCSS **K.OA.1, K.OA.2, K.OA.5, SMP1, SMP6, SMP7**

## Get Yourselves in Order

### Counting and Cardinality

Ask children to raise their hands. Then have them count off as you point to them at random. As they give the next number in order, have them put down their hands. Tell children to remember their numbers.

When all children have received numbers and all hands are down, have them line up in numerical order.

CCSS K.CC.1, K.CC.2, K.CC.4

# Greater or Less

## Counting and Cardinality

Ask children to explain or "prove" their answers to the prompts below. Display the numbers if possible.

*Is 2 greater than 6, or is it less than 6?*

*Is 14 greater than 19, or is it less than 19?*

*Is 54 less than 55, or is it greater than 55?*

*Is 23 less than 28, or is it greater than 28?*

*How do you know?*

*Is it true that 42 is less than 43? Is it true that 33 is less than 35? Is it true that 27 is less than 22? Is it true that 50 is greater than 48? How do you know?*

 CCSS K.CC.1, K.CC.2, K.CC.7, SMP3

## Quick Facts

### Operations and Algebraic Thinking

Encourage children to solve these (and similar "within 5" addition and subtraction problems) using the quickest strategy they are comfortable with:

*4 plus 1 makes how much?* (5)

*What does 1 plus 2 equal?* (3)

*What is 6 minus 2?* (4)

CCSS **K.OA.1, K.OA.2, K.OA.5, SMP6**

# About How Many?

## Counting and Cardinality

Ask: *About how many chairs do you think there are there in our classroom? More than 10? Fewer than 10? More than 50? More than 100?*

*About how many tables?*

Have children count to check their estimates and compare them to the actual counts.



CCSS **K.CC.4, K.CC.5, K.CC.6, K.CC.7, SMP6**

## What Number Comes Before?

### Counting and Cardinality

Ask: *When you count, what number comes right before 20? Before 14? Before 16? Before 10? Before 13? Before 15? Before 21? Before 49?*

## Share a Number Story

### Operations and Algebraic Thinking **Number Stories**

Invite a child to share an 8 story for the class to solve. (For example: *Four bears are in a cave. Four more bears join them. How many bears in all?*)

Then invite another child to tell a different 8 story. Compare the stories.

Repeat with other stories, such as a 4 story or a 9 story.



CCSS **K.OA.1, K.OA.2, K.OA.3, SMP1**

## Comparing Numbers

### Counting and Cardinality

Present these pairs of numbers orally or display them if needed.

*Which number is greater: 13 or 15? 17 or 14? 61 or 62? 52 or 50? 40 or 30? 23 or 33? 6 or 16?*

*Which number is less: 13 or 15? 16 or 18? 29 or 19? 10 or 20? 40 or 14? 12 or 20?*

Ask: *How do you figure out whether a number is greater or less? Does that work for all numbers? Does anyone do it a different way?*

CCSS **K.CC.1, K.CC.2, K.CC.3, K.CC.7, SMP3**

# Counting Down—Teens

## Counting and Cardinality

Use a number line for the following if necessary.

Say: *Give the numbers in order from 14 to 12; 13 to 10; 17 to 15; 16 to 14; 19 to 17; 15 to 13; 18 to 14; 20 to 10;* and so on.



**CCSS K.CC.1, K.CC.2, K.CC.3, SMP5, SMP6**

# Clothing Combinations

## Operations and Algebraic Thinking **Number Stories**

Mariah packed shorts and T-shirts for her trip. There were 6 pieces of clothing total. What are the different combinations of clothing that she might have had in her suitcase? (1 pair of shorts and 5 T-shirts; 2 pairs of shorts and 4 T-shirts; and so on)

Austin had striped and plain-colored shirts in his drawer. There were 8 shirts total. What are the different combinations of shirts he might have had in his drawer? (1 striped and 7 plain; 2 striped and 6 plain; 3 striped and 5 plain; 4 striped and 4 plain; and so on)

CCSS **K.OA.1, K.OA.2, K.OA.3, SMP1**

# More Standing in Line

## Counting and Cardinality

When children are in line, ask them to count off out loud and to remember their numbers. Then ask: *Who is ninth in line? 13th? Fourth? Seventh? 21st? What would happen if we reversed the order of our line?*



CCSS K.CC.1, K.CC.4

## "I'm Thinking of a Number"

### Counting and Cardinality

Say: *I'm thinking of a number that is less than 20. It's more than 10. It equals 10 plus 3 more. What is it?* (13) Give one clue at a time until a child guesses the answer. Then ask children to give other clues that would fit the number.

Repeat with other numbers, focusing on the range between 0 and 20.

CCSS **K.CC.7, K.OA.1, K.OA.2, K.OA.3, SMP1**

# Think of a Number Story

## Operations and Algebraic Thinking

Number Stories

Invite a child to share a 10 story for the class to solve.
(For example: *There were 20 children in class. Then 10 children went to lunch early. How many children are left in class?*)

Then invite another child to tell a different 10 story.
Compare the stories.

Repeat with other stories, such as a 0 story or a 7 story.



CCSS K.OA.1, K.OA.2, K.OA.3, SMP1

# More Simple Addition

## Operations and Algebraic Thinking

Say: *Solve these using the quickest strategy you are comfortable with:*

*1 + 1*

*3 + 1*

*4 + 1*

*7 + 1*

*What strategy did you use? What do you notice? What might we call your strategy(ies)?*

CCSS **K.OA.1, K.OA.2, K.OA.5, SMP1, SMP6, SMP7**

## About How Many?

### Counting and Cardinality

Ask: *About how many children here have brown hair? About how many have black hair? Are your estimates closer to 10 or closer to 100? Why?*

*About how many children do you think are wearing blue today?*

After the children have had an opportunity to make and explain their guesses, count to verify. If possible, display the estimates and the actual count for comparison.



**CCSS K.CC.4, K.CC.5, K.CC.6, K.CC.7, SMP6**

# Name 3-Dimensional Shapes

## Geometry

Ask children to name some 3-dimensional shapes they know. (Cube, cone, cylinder, sphere, and so on)

Then ask: *Do you see any of these shapes in our classroom? Where?*

*What's the same about all of the cylinders* (or another shape) *you see in the classroom? What's different about them?*

CCSS K.G.1, K.G.2, K.G.4, SMP6, SMP7

## Apple Math

### Operations and Algebraic Thinking Number Stories

Remember to use your children's names in number stories like the following:

*Jeffrey brought a box of apples to school. It weighed 8 pounds. The box by itself weighed 1 pound. How much did the apples weigh?* (7 pounds) *How did you figure it out?*

*There were apple slices for snack that day. Yoshi, Alix, and Joan sat at a table together. They were given 6 apple slices. If the 3 girls shared the apple slices equally, how many did each girl get?* (2 slices) *How did you figure it out?* (You may want the children to act this out.)



CCSS K.CC.6, K.OA.1, K.OA.2, SMP1, SMP2, SMP4

# Fair Shares

## Operations and Algebraic Thinking

Number Stories

Use your children's names and have children act out number stories like the following.

*Ann has 4 stuffed animals. She shares them equally with Fred. How many stuffed animals do they each have?* (2 stuffed animals) *How did you figure it out?*

*Tomi has 6 marbles. He shares them equally with Ellen. How many marbles do they each have?* (3 marbles) *How did you figure it out?*

CCSS K.OA.1, K.OA.2, SMP1

# Counting Shoes by 2s

## Counting and Cardinality

Say: *Let's count how many shoes are in our class. A quick way to count them is to count by 2s.*

Have the children stick out their feet. Have children listen and join in as you count the feet by 2s.



CCSS K.CC.1, K.CC.5, SMP7

# Solving 2-Step Problems

## Operations and Algebraic Thinking

**Number Stories**

Ken had 2 marbles. He won 2 more and then lost 3. How many marbles did he have at the end? (1 marble) How do you know?

Nomi's dog had 6 puppies. Nomi gave 2 puppies away, but then got 1 of them back. How many puppies does Nomi have now? (5 puppies) How do you know?

CCSS **K.OA.1, K.OA.2, SMP1**

# Compare Areas

## Measurement and Data

Point to two rectangular objects that are obviously different in size (the board and a desk; a book and a window). Ask children which object they think would take more paper to cover.

(Children who have doubts can check it out later.)



CCSS K.MD.1, K.MD.2, SMP1

# Reading Class Graphs

## Measurement and Data

Ask children to answer questions about graphs made previously in the classroom.

For example, using the Types of Pets Graph (Lesson 6-3), ask:

*How many children have dogs?*

*Which type of pet do more children have than any other type of pet?*

*How many more children have cats than dogs?*

*How did you use the graph to figure out your answers?*

CCSS **K.CC.5, K.CC.6, K.MD.3, SMP1, SMP4**

## How Many All Together?

### Operations and Algebraic Thinking **Number Stories**

Maurice has 3 yellow pencils, 4 purple ones, and 1 blue one.
How many pencils does he have in all? (8 pencils) How do you know?

In a pet show, there are 3 dogs, 5 cats, and 1 hamster.
How many pets are there in all? (9 pets) How do you know?

 CCSS K.OA.1, K.OA.2, SMP1

# Telling Number Stories to Fit Equations

## Operations and Algebraic Thinking **Number Stories**

Give the following prompts:

*Tell us a 2 + 3 story.*

*Tell us a 5 − 1 story.*

*Tell us a 4 + 2 story.*

Provide examples if needed, such as the following stories for 2 + 3:

*There were 2 picture books and 3 chapter books on the shelf. How many books were there all together?*

*Two children arrived at school by walking. Then 3 more children got off the school bus. How many children were at school then?*

CCSS K.OA.1, K.OA.2, SMP1, SMP4

# Tell Numbers After and Before

## Counting and Cardinality

Ask: *When you count, what number comes after 25? After 69? After 99?*

Then say: *Tell me any number that comes before 51; before 44; before 11; before 90.*



CCSS K.CC.1, K.CC.2

# Simple Subtraction

## Operations and Algebraic Thinking

Say: *Solve these using the quickest strategy you are comfortable with:*

*5 − 0*

*2 − 0*

*4 − 0*

*What strategy did you use? What do you notice? What might we call your strategy(ies)?*

## Temperature Change

### Operations and Algebraic Thinking

Number Stories

Demonstrate degrees on a thermometer. Ask: *How much does the temperature rise when it goes from 70 degrees to 76 degrees?* (6 degrees)

*The thermometer read 50 degrees at noon. It was 10 degrees colder (lower) at night. What did the thermometer read then?* (40 degrees)



CCSS K.OA.1, K.OA.2, SMP1

## How Far? How Much?

### Operations and Algebraic Thinking **Number Stories**

For each of these (or similar) problems, ask children to share and explain how they figured out their answers.

*Marta walked 6 blocks to school, and Leslie walked 3 blocks less than Marta. How many blocks did Leslie walk?* (3 blocks)

*The rug in Kendall's room is 8 feet wide. It is 2 feet longer than it is wide. How long is her rug?* (10 feet long)

*Jason weighs 42 pounds. Glen weighs 1 pound less than Jason. How much does Glen weigh?* (41 pounds)

CCSS **K.OA.1, K.OA.2, SMP1**

# What Number Follows?

## Counting and Cardinality

Ask: *When you count, what number comes after 75? After 86? After 52?*

*When you count, what number follows 23? 16? 47? 64? 99?*

 CCSS K.CC.1, K.CC.2

## More Quick Facts

### Operations and Algebraic Thinking

Encourage children to solve these using the quickest strategy they are comfortable with:

*How much is 5 minus 3?* (2)

*What does 1 plus 4 equal?* (5)

*What does 3 plus 2 make?* (5)

# Counting Dimes by 10s

## Counting and Cardinality

Explain to children that a dime is worth 10 cents. If possible, use real or play dimes as you count.

Ask: *How many cents do I have if I have 6 dimes? Let's count by 10s to find out. 10 cents, 20 cents, 30 cents, . . . and so on.*

Repeat with other quantities of dimes (1–10).



CCSS K.CC.1

# Which Takes Longer?

## Measurement and Data

Ask: *Which would take longer to sweep, the Math Center or the classroom? Why do you think so? How could we find out?*

# "I'm Thinking of a Shape" (3-Dimensional)

## Geometry

Give clues about 3-dimensional shapes and have children narrow down their choices by guessing.

*I'm thinking of a 3-dimensional shape with flat sides. What could it be? What else might you want to know to figure it out?*

*I'm thinking of a 3-dimensional shape with a round face. What could it be? What else might you want to know to figure it out?*



CCSS K.G.2, K.G.4, SMP7

# Name Numbers Before, After, and Between

## Counting and Cardinality

Say: *Name the numbers before and after 15; 28; 50; 73; 100;* and so on.

Then ask: *What numbers come between 56 and 64? Between 13 and 8? Between 37 and 42? Between 89 and 92?*

## Lunch Math

### Operations and Algebraic Thinking **Number Stories**

For each of these (or similar) problems, ask children to share and explain how they figured out their answers.

*In her lunch box, Tori has 4 carrot sticks and 3 celery sticks. She has as many crackers as she does vegetable sticks. How many crackers does she have?* (7 crackers)

*Tom had 4 pieces of cheese in his lunch and ate 3 of them. Then he gave 1 piece of cheese to Jonathan. How many pieces of cheese does Tom have left?* (0 pieces)



CCSS **K.OA.1, K.OA.2, SMP1**

## More Quick Facts

### Operations and Algebraic Thinking

Encourage children to solve these quickly:

*What does 3 plus 2 equal?* (5)

*What does 5 minus 3 equal?* (2)

*How much is 2 plus 3?* (5)

*Do you notice any connections between those three problems?* (Children may notice the reciprocal relationship between $3 + 2 = 5$ and $5 - 3 = 2$, or that $3 + 2$ equals the same thing as $2 + 3$.)

**CCSS K.OA.1, K.OA.2, K.OA.5, SMP6, SMP7**

## Shopping for 10-Cent Items

**Operations and Algebraic Thinking.** Number Stories

Ask children to pretend that they are in a 10-Cent Store. Encourage them to count by 10s.

*Tami wants 3 kites. They cost 10 cents each. How much must she pay?* (30 cents)

*How many 10-cent stickers can Connie buy for 40 cents?* (4 stickers)

*Each marker costs 10 cents. How much will 9 markers cost?* (90 cents)

 CCSS K.CC.1, SMP1

## Counting through Decade Changes

### Counting and Cardinality

Have the children repeat after you: *9, 10, 11; 19, 20, 21; 29, 30, 31; 39, 40, 41; 49, 50, 51; 59, 60, 61; . . . 99, 100, 101.* Explain that this is a way to help them practice the trickiest parts of counting to 100.

Practice counting on from different starting points through these decade changes during downtimes throughout the day.

As a variation, have children sit in a circle and go around with each child supplying a missing number.

CCSS **K.CC.1, K.CC.7, SMP8**

# Switching Numbers

## Counting and Cardinality

Say: *I will name some numbers in a string. I will switch 2 of the numbers around. Which numbers are switched?*

*24, 25, 27, 26, 28, 29* (27 and 26)

*20, 19, 17, 18, 16, 15* (17 and 18)

*4, 5, 7, 6, 8* (7 and 6)

*9, 7, 8, 6, 5* (7 and 8)



CCSS K.CC.1, K.CC.2, K.CC.4, SMP6

# More Simple Subtraction

## Operations and Algebraic Thinking

Say: *Solve these using the quickest strategy you are comfortable with:*

*4 − 1*

*5 − 1*

*3 − 1*

*What strategy did you use? What do you notice? What might we call your strategy(ies)?*

CCSS K.OA.1, K.OA.2, K.OA.5, SMP1, SMP6, SMP7

# Lemonade Stories

## Operations and Algebraic Thinking Number Stories

One pitcher contained 4 cups of lemonade and another contained 3 cups. How many cups of lemonade were there in all? (7 cups)

Children drank 5 of the cups of lemonade at lunch. How many cups of lemonade were left? (2 cups)

# At the Park

## Operations and Algebraic Thinking

Number Stories

Two boys raced down the hill. Adam got to the bottom in 30 seconds, but Ariel arrived in 20 seconds. Who got to the bottom first? (Ariel) How long did Ariel wait for Adam? (10 seconds) How did you figure it out?

CCSS K.CC.1, K.OA.1, K.OA.2, SMP1

# Skip Counting

## Counting and Cardinality

Begin counting by 10s (or 5s or 2s). Stop counting and point to a child, who then says the next numbers in sequence. Stop the child with a signal (a stop sign or your raised hand) and point to another child, who continues counting.

Keep counting and stopping until you reach a desired number.



CCSS K.CC.1, K.CC.2, SMP7

# How Many of a Kind?

## Operations and Algebraic Thinking

Number Stories

A mother hen has 7 chicks, and 5 of these chicks are black. The others are yellow. How many chicks are yellow? (2 chicks) How do you know?

A mother cat has 6 kittens, and 2 of these kittens are spotted. The others are gray. How many kittens are gray? (4 kittens) How do you know?

CCSS K.OA.1, K.OA.2, SMP1

# Find Numbers After and Before

## Counting and Cardinality

Ask: *When you count, what is the number that comes after 42? After 76? After 23? After 100?*

Then say: *Tell me the number that comes right before 56; before 68; before 37; before 90; before 100.*



CCSS K.CC.1, K.CC.2

## Sharing

### Operations and Algebraic Thinking Number Stories

For each of these (or similar) problems, ask children to share and explain how they figured out their answers.

*If Terrance took 10 books and put them into 2 equal piles, how many books would be in each pile?* (5 books)

*Mia and Jason shared 6 strawberries. Mia ate 1 and Jason ate 5. How else could they have shared the strawberries?* (2 and 4; 3 and 3; 4 and 2; and 5 and 1)

CCSS K.OA.1, K.OA.2, K.OA.3, SMP1

# What Do You Add or Subtract?

## Operations and Algebraic Thinking

Encourage children to listen carefully and respond.

*What number do you subtract from 8 to get 8?* (0)

*What number do you add to 1 to get 5?* (4)

*What number do you subtract 1 from to get 4?* (5)

*What number do you subtract 2 from to get 1?* (3)



CCSS K.OA.1, K.OA.2, K.OA.5, SMP1

# How Many Left?

## Operations and Algebraic Thinking Number Stories

For each of these (or similar) problems, ask children to share and explain how they figured out their answers.

*The children saw 8 penguins on a rock at the zoo. Soon 4 large penguins and 3 baby penguins slid back into the water. How many of the penguins were left on the rock?* (1 penguin)

*Louisa folded 2 origami bunny rabbits and 2 origami ducks. She gave 3 of the origami animals to Tina. How many did she have left?* (1 animal)

CCSS K.OA.1, K.OA.2, SMP1

# How Much Do You Add or Subtract?

## Operations and Algebraic Thinking

Encourage children to listen carefully.

*How much do you add to 2 to get 4?* (2)

*How much do you subtract from 4 to get 3?* (1)

*How much do you add to 1 to get 5?* (4)



CCSS **K.OA.1, K.OA.2, K.OA.5, SMP1**

# What Number Comes After? (Large Numbers)

## Counting and Cardinality

Ask: *When you count, what number comes after 33? After 50? After 75? After 102? After 58?*

*When you count, what number follows 64? 72? 101? 47? 13?*

CCSS **K.CC.1, K.CC.2**

## Magician Tricks

### Operations and Algebraic Thinking **Number Stories**

Jan's uncle is a magician. He put 1 scarf in his top hat. Jan reached in and found 2 scarves. Then the magician put 2 rubber balls in his hat. Jan reached in and found 3 rubber balls. If the magician puts 4 bananas in his hat, how many do you think Jan will find there? (5 bananas) Why?

What happens each time Jan reaches into the hat? (Jan finds 1 more thing has been added.)



CCSS K.CC.4, K.OA.1, K.OA.2, SMP1, SMP7, SMP8

## Sharing Food

### Operations and Algebraic Thinking

Number Stories

Allow children to act out these stories as needed. If they use other strategies, encourage them to share and explain them.

*There are 10 muffins in a box. Half of the muffins are blueberry. How many muffins are blueberry?* (5 muffins)

*Four children shared a box of raisins equally. There were 8 raisins in the box. How many raisins did each child take?* (2 raisins)

CCSS K.CC.5, K.CC.6, K.OA.1, K.OA.2, SMP1, SMP2

# Counting On—3 Numbers

## Counting and Cardinality

Say: *Counting on, name the 3 numbers that follow 76; 44; 88; 92; 98;* and so on.



CCSS K.CC.1, K.CC.2

## Sibling Stories

### Operations and Algebraic Thinking

Number Stories

For each of these (or similar) problems, ask children to share and explain how they figured out their answers.

*Zach is 6 years old. His big brother, Jake, is 3 years older. How old is Jake?* (9 years old)

*Zach and Jake have a younger brother named Colin who is 2 years younger than Zach. How old is Colin?* (4 years old)

*Jake, Zach, and Colin went to the rodeo on Saturday. They saw 7 white horses and 3 black horses. How many more white horses did they see than black horses?* (4 more white horses) *How many horses did they see all together?* (10 horses)

CCSS K.OA.1, K.OA.2, SMP1

# What Do I Have to Do?

## Operations and Algebraic Thinking

Say:

*I have 5. I want 7. What do I have to do?* (Add 2.)

*I have 6. I want 3. What do I have to do?* (Subtract 3.)

*I have 10. I want 8. What do I have to do?* (Subtract 2.)



CCSS K.OA.1, K.OA.2, SMP1, SMP2

# Buying with 10 Cents

## Operations and Algebraic Thinking

Number Stories

Nick wants to buy a 10-cent ticket. He has only 7 cents. How many more cents does he need? (3 more cents)

Jared has 8 cents. He wants to buy a toy car that costs 10 cents. How many more cents does he need? (2 more cents)

Daniella had 10 cents. She bought a paintbrush for 6 cents. How many cents does she have left? (4 cents left)

If you have 10 cents and spend it all on 2 balls that cost the same amount, what does each ball cost? (5 cents each)

CCSS K.OA.1, K.OA.2, K.OA.4, SMP1, SMP6

# More Find Numbers Between

## Counting and Cardinality

Say: *Name 5 different numbers that are greater than 58 and less than 64.*

*Name 5 different numbers that are greater than 84 and less than 90.*



CCSS **K.CC.1, K.CC.2, K.CC.7**

# Simple Addition and Subtraction

## Operations and Algebraic Thinking

Say: *Solve these using the quickest strategy you are comfortable with:*

*4 − 0*

*2 + 1*

*3 + 0*

*5 − 1*

*1 + 1*

*3 − 1*

*2 + 0*

*5 − 0*

*What strategies did you use? What do you notice?*

CCSS **K.OA.1, K.OA.2, K.OA.5, SMP1, SMP6, SMP7**

## How Many More Do I Need?

### Operations and Algebraic Thinking

Number Stories

Suggest that children count on and keep track with their fingers if needed to solve the following problems:

*I have 5 cups. I need 8 cups for my party. How many more cups do I need?*
(3 more cups)

*I have 9 eggs. How many more do I need to make a dozen (12)?*
(3 more eggs)



CCSS K.CC.2, K.OA.1, K.OA.2, SMP1, SMP2

## Just Before Lunch

### Counting and Cardinality

Ask: *How long do you think it will take us to wash and line up for lunch? Let's time ourselves by counting with a steady beat while we get ready.*

*How many counts did it take us? What do you think our count would have been if we had counted faster? Slower? Why?*

CCSS K.CC.1, SMP6

## Which Takes Longer?

### Operations and Algebraic Thinking

Number Stories

Sparrow eggs hatch in 2 weeks. Penguin eggs hatch in 8 weeks. Which eggs take longer to hatch? (Penguin eggs) How much longer do they take? (6 weeks longer) How did you figure out your answers?



CCSS K.CC.7, K.OA.1, K.OA.2, SMP1

# Counting from 90 to 115

## Counting and Cardinality

Say: *Give the numbers right before and after 100; 110; 114;* and so on.

*Let's count from 90 to 115 together.*

CCSS K.CC.1, K.CC.2

# More Telling Number Stories to Fit Equations

## Operations and Algebraic Thinking Number Stories

Give the following prompts:

*Tell us a 9 + 1 story.*

*Tell us a 9 − 1 story.*

*Tell us an 8 + 2 story.*

Provide examples if needed.

 CCSS K.OA.1, K.OA.2, SMP1, SMP4

## Which Is Faster?

### Counting and Cardinality

Ask: *Do you think it would be faster to crawl across the room or walk across the room? Why?*

*If we count with a medium, steady beat, do you think it will take closer to 10 counts, 50 counts, or 100 counts to walk across the room? Let's count and see!*

CCSS **K.CC.1, K.MD.2, SMP3**

## Operations and Algebraic Thinking **Number Stories**

In the story, "Jack and the Beanstalk," Jack wondered how high his beanstalk grew. If you were Jack, how would you measure it?

If Mama and Papa Bear from the story, "The Three Bears," have 2 more cubs after Baby Bear, how many cubs will be in the bear family? (3 cubs) How can we show this with a number sentence? ($1 + 2 = 3$) How many bears will be in the bear family all together if Mama and Papa Bear have 3 cubs? (5 bears) How can we show the total number of bears with a number sentence? ($2 + 3 = 5$ or $2 + 1 + 2 = 5$)

In the story, "The Three Little Pigs," how many bricks do you think the third little pig used to make his house? Fewer than 10? More than 10? More than 50? More than 100? Much more than 100? Why do you think so?

 CCSS **K.OA.1, K.OA.2, SMP1, SMP5, SMP6**

## More Teen Finger Math

### Number and Operations in Base Ten

Hold up 10 fingers.

Have a partner hold up fingers next to yours to make 12; then 15; then 18.

Ask: *How do you know how many to hold up?*

CCSS **K.NBT.1, SMP2**

## Which Number Is Smallest?

### Counting and Cardinality

Say: *Tell me which number is the smallest.*

*27, 26, 25, 24, 23* (23)

*22, 23, 24, 25, 26* (22)

Next, begin with 3 numbers in mixed order. Increase the length of the number strings as children catch on.

*This time I'm going to be tricky. Listen closely and tell me which number is the smallest.*

*27, 22, 26* (22)



**CCSS K.CC.1, K.CC.2, K.CC.7, SMP1**

# What Do I Add to Make 10?

## Operations and Algebraic Thinking

Encourage children to listen carefully and respond:

*What do I add to 5 to make 10?* (5)

*What do I add to 9 to make 10?* (1)

*What do I add to 7 to make 10?* (3)

*What do I add to 6 to make 10?* (4)

*What do I add to 2 to make 10?* (8)

CCSS K.OA.1, K.OA.2, K.OA.4, SMP1, SMP7

# Tell Me about 3-Dimensional Shapes

## Geometry

Say: *We are going to tell each other about 3-dimensional (solid) shapes.*

*I'll tell you about a cube. It has 6 faces that are all squares.*

*Now you tell me about a sphere; a cone; a cylinder; a rectangular prism.*

 CCSS K.G.2, K.G.4, SMP6, SMP7